# WHY DOES CLASSICAL MECHANICS FORBID INERTIAL PROPULSION DEVICES WHEN THEY EVIDENTLY DO EXIST?

# A MAJOR REVISION OF THE AUTHOR'S POINT OF VIEW BASED UPON PROF. WALTER NOLL'S OBSERVATIONS PRIOR TO HIS AXIOMATIZATION OF CONTINUUM MECHANICS

## DENNIS P. ALLEN, JR.

# DEDICATION

This book is respectfully dedicated to the Holy Spirit of God, Source of all Wisdom and Knowledge and the Spirit of Truth, together with His Most Chaste Spouse, the Blessed Virgin Mary, without Whom this book could have neither been conceived nor written.

However, any and all mistakes are, of course, solely the author's responsibility.

# CONTENTS

DENNIS P. ALLEN, JR.

# ACKNOWLEDGEMENTS

The author would like once again to express thanks to his physics mentor, (the late) Thomas E. Phipps Jr., PhD (Harvard, 1951), for patiently explaining to him (at length) that lower order physics must be made right before higher order physics can be made right.  Also, thanks to Greg Volk for his help, encouragement, and assistance.  And the author owes a vote of thanks to Harvey Fiala and his explanation of his HMT working gyroscopic inertial propulsion device during our many phone conversations.  And many thanks to Nick Percival for his suggestions as to how to make our work clearer and more correct.  Thanks also to the reference librarians at the Spring Lake District Library for their assistance in preparing this manuscript.  And the author would like to thank Prof. James Casey for recently pointing out some errors in an earlier version to him.  Finally, the author would like to thank Gottfried Gutsche for his free copies of his inertial propulsion books that then succeeded in convincing him that there definitely *are* difficulties with Newton's third law of motion ... and that proper use of energy methods *may* be a key to avoiding these problems.

DENNIS P. ALLEN, JR.

# INTRODUCTION

Inertial propulsion is the ability to move linearly and indefinitely a device either in three dimensional space or on a two dimensional surface using no propellers, exhaust gasses, or traction against a surface (say the device is held by gravity to the surface), but only using the internal dynamics of this device. And it is considered by classical mechanical experts as not existing ... since it manifestly violates the separate conservation of angular and linear momentum. However, there are numerous working devices invented by such people as Harvey Fiala (a retired Space Shuttle Engineering Reentry Supervisor), Gottfried Gutsche (a semi-retired German engineer and author), and Veljko Milkovic (a Serbian inventor and author) that may be viewed in operation on the Internet (see his excellent web site) or in videos of talks given at conferences. [See on the web site "ResearchGate" under the author's name a lengthy computer simulation of the Milkovic oblique pendulum driven cart; but actually, for this to qualify as a bona fide inertial propulsion device, one has to continue on and imitate Christiaan Huygens who studied simple pendulums' dynamics and who invented the simple pendulum clock by adding an escapement to it to give it a "boost" at the end of each cycle so that the bob would not quickly run down.]

Gottfried Gutsche has published a series of books (available from Amazon.com) on inertial propulsion such as [1]. In them, he describes some patented inventions using classical mechanical formulas from the very well-known Kurt Gieck Engineering Formulas 7th Edition-section L1-L10. One of these inventions is an inertial propulsion device called the MARK II Inertial Propulsion Device, a working model of which he demonstrates on the Internet, and which is *not* a gravity machine (and so should then also work in free fall) as are Fiala's HMT and Milkovic's oblique pendulum driven cart. He also has invented other devices which illustrate the superiority of energy methods to (ordinary) momentum methods (where the force is generally the time derivative of the momentum).

Additionally, Harvey Fiala has a HMT gyroscopic inertial propulsion device (a gravity machine) that he discusses and demonstrates in operation in his 2012 TeslaTech talk, a video of which may be purchased from them [3].

Actually, Dr. Jeremy Dunning-Davies and the author have written a book [2] that allows the analysis of Fiala's HMT (as he calls it) and gyroscopic devices in general from the point of view of changing inertial mass, but our work shows that rotor mass only changes appreciably in the case of very high rotation; and so, since the Milkovic cart (mentioned above) has only a relatively slowly swinging pendulum (in an oblique plane), our changing inertial mass notions would *not* allow an analysis of this cart that was appreciably different than a classical mechanical one. Thus we realized that even in the case of Newtonian mechanics holding, there are other problems than just changing inertial mass ones. And, unfortunately, the Fiala HMT device also finally proved to have its gyroscope rotor spinning just too slowly for an appreciably different Neo-Newtonian analysis. But we *do* have a completely worked out example (see ResearchGate.com under the author's name) of an elementary hoop rotor (with massless spokes) gyroscope having a fixed pivot point that is steadily & horizontally precessing (i.e. without any nutation) and that is assumed to have its rotor spinning with a (high) angular velocity of 388 Hertz which has a Neo-Newtonian analysis that *is* appreciably different than the corresponding Newtonian one.

See also our new Appendix 4 concerning long time Boeing engineering supervisor Michael Gamble's recent COFE7 talk chronicling Boeing's long history of using "Control Moment Gyros" (that is, inertial propulsion of the forced precession type) to alter their satellites' orbits without the burning of *very* expensive propellant.

And we have now additionally added an introduction to Gottfried Gutche's mechanical writings found in his various inertial propulsion device books as our Appendix 5 … in as much as his thinking so far has proven to be quite opaque both to mechanical engineers and physicists interested in classical mechanics.

We would like to thank Prof. James Casey (of the Univ. of California at Berkeley's Mechanical Engineering Dept.) for sending a proof of the conservation of linear momentum from H. Lamb's *Dynamics* [10], and for finding an error in the author's supposed example of a three particle counter-example to the consistency of Newton's third law and his first two laws. And then Prof. Casey suggested that the author read Prof. Walter Noll's 1957 axiomatization of classical mechanics that gave the author the key idea for explaining the failure of conservation of momentum in the case of the Milkovic oblique pendulum driven cart, namely, the exiting of the realm of classical mechanics and the entering of the realm of quantum mechanics in the cart-earth system.

Finally, we note that our argument that inertial mass in Newtonian mechanics is variable even at non-relativistic velocities has been found to be incorrect by Prof. Dr. Chris Provatidis, and due to an erroneous implicit approximation involving a moment of inertia. The author thanks him for pointing this out to him.

# WHY DOES CLASSICAL MECHANICS FORBID INERTIAL PROPULSION DEVICES WHEN THEY EVIDENTLY DO EXIST?

# 1 Our New Point Of View

In our previous work on inertial propulsion devices, we laid their non-conservation of momentum at the door of the failure of Newton's third law of motion since both Heinrich Hertz and [12] Gottfried Gutsche [1] mention that this law of motion seems out of step with the first two laws of motion. However, Prof. James Casey of the Univ. of Calif. at Berkeley's Mechanical Engineering Dept. kindly sent the author several emails in which he delved deeply into his reasoning and finally convinced the author that Newton's third law IS consistent with his first two laws (although further experimentation could still show it to be incorrect … as the last experiment is never done).

But if it is actually true that the Veljko Milkovic oblique pendulum cart (see Appendix 7 below for a complete description of this cart) involves the non-conserving of momentum (and we adduce hard evidence that this is so in our cart computer simulations posted on our ResearchGate.com web page) and since the moving parts of this device are observed to move too slowly for either Eric Laithwaite's "mass transfer" or his "variable inertial mass" to play an important role; then how can a computer simulation of this cart using only Newton's three laws with and the other parts of his mechanics (such as his parallelogram rule for adding forces) -- but without using any conservation of momentum – give good results?

Well, the usual explanation of this is that when we view this cart in motion, we are actually viewing a very complicated "differential friction effect" of the sort of where the cart has a greater rolling wheel friction force when moving backwards than moving forwards so that momentum (properly understood) *is* actually conserved.    However, when the coefficients of friction are all set to zero (a condition that can very closely be realized with the new high-tech bearings and in a vacuum chamber) and the simulation of the cart is then redone without  any further significant changes, the inertial propulsion does not cease (as it would be expected to by the friction effect naysayers), but instead becomes even more robust!

In the following Chapter 2, we will analyze these matters in detail.

DENNIS P. ALLEN, JR.

# 2. Re-Analyzing The Phenomenon

So, then, if one accepts that Newton's mechanics DOES hold true in the case where one has a (small) inertial propulsion device with slow moving parts, how can this apparent non-conservation of momentum happen in a situation where its inertial propulsion can be shown not to fundamentally involve any kind of frictional effect (as with the Milkovic oblique pendulum driven cart)?

Well, the author would like to thank Prof. James Casey of the Univ. of Calif. at Berkeley's Mechanical Engineering Dept. for referring him to an axiomatization of classical mechanics due to a very accomplished Mechanical Engineering professor, Walter Noll (recently deceased), that appeared in [7] in 1957. In this important paper (see Appendix 6 for a reproduction of the first page of this article that gave the author his key idea), Noll advances the idea that the difficulty of basing continuum mechanics upon Newton's mechanics with a larger and larger number of smaller and smaller particles resides in the exiting of the realm of classical mechanics and the entering the realm of quantum mechanics that is, of course, governed by quite different laws. In fact, Noll begins by stating (again, see attached page 266 of Appendix 6) that it is *not* true that continuous bodies can be regarded as the limiting case of a classical particle system with an increasing number of particles ... although they *can* in the elementary special cases of "perfect fluids" and "linearly elastic solids". And the reason Noll gives is that, in the general case of continuum mechanics, this larger and larger particle approach involves the leaving of the classical mechanical realm and the entering of the quantum mechanical realm which, of course, is described by quite different laws ... the connection between these two realms being the "weak" link of statistical mechanics.

So, then, Noll is apparently saying that, if during one's classical mechanical calculations and deductions, one wanders off into too small of particles and so into the realm of quantum mechanics, then one cannot reasonably expect to obtain the correct solutions to one's real world mechanical problems ... even if one has wandered off unknowingly, of course.

With this in mind, it seems to the author that the Milkovic cart is yet another example of where classical mechanics also breaks down when it insists that the usual separate conservation of linear and angular momentum can be safely applied to the cart-earth system in order to show that the motion of the cart – when started with the oblique pendulum's shaft being horizontal and with the cart being on a smooth and level surface – is thus merely a rather complicated "differential friction" effect ... or at least *some* kind of frictional effect. And to check this, the author has rerun his computer simulation of the Milkovic cart, but with the two coefficients of friction both zeroed so that the cart is then (assuming no air friction as in a vacuum) frictionless; and it then behaves in much the same way as before (with 2% rolling wheel friction and small pendulum bearing friction) except that (1) there is more movement with no friction (as expected) and (2) there is, of course, energy conservation without friction so that, after some brief transient behavior, it exhibits cyclic behavior ... and so never runs down or stops translating. Now, of course, this "no friction" situation cannot be exactly checked experimentally; but with high-tech bearings it can be checked (in a vacuum chamber) to a close approximation.

Now, Noll uses "mass" as a primitive notion in his axiomatization of continuum mechanics while Prof. Herbert Simon uses time, position, velocity, and acceleration only. Well, Simon's approach to the axiomatization of classical mechanics is an "operational" approach, where "operationalism" comes from Prof. P.T. Bridgeman's very, very influential book, "The Logic of Modern Physics". This approach targets the measurement of the various physical concepts such as mass, charge, time, position, velocity and acceleration. And he notes that mass is *much* more complicated than Simon's just mentioned four primitive notions both to define and then to measure. And, of course, this is largely because of relativistic mass gain involving, then, Einstein's SRT.

But a mechanical engineer might well object here that he does *not* consider relativistic velocities, and should he want to measure the mass of an object, he simply places it upon a scale! However, this procedure involves the equality of gravitational and inertial masses ... which Newton thought he had shown via some pendulum experiments. But the last experiment is never done, and recently Prof. Alexander Dmitriev of St. Petersburg, Russia, has done some very clever falling rotor experiments [8] with horizontally spinning rotors that he has found do *not* fall with the same accelerations when spinning and not spinning ... with this spin falling acceleration discrepancy being a function of their spin angular velocity.

Now, if one assumes that the rotor inertial masses are a function of their spin here, then one can formulate a new mechanics which is pretty much everywhere the same as classical mechanics except that particle inertial mass is variable ... as in the case of a rocket whose mass decreases with the burning of its propellant. And so Dr. Jeremy

Dunning-Davies and I have authored a two volume book series, "Neo-Newtonian Mechanics with Extension to Relativistic Velocities", the first book presenting this Neo-Newtonian Mechanics in its Chapter 4 and its appendix. Thus, even just within the realm of classical mechanics, it certainly *does* appear that Bridgeman's thinking that mass is a much more complicated notion than those four primitive notions of Simon's axiomatization of classical mechanics and therefore should *not* be deemed a primitive notion (as Noll would suggest was valid) from the operational point of view, that is, from the point of view of its measurement by *equipment*, is correct.


And, since if one considers the cart-earth system, then a typical sheer force against the tracks that the cart is assumed to be on (which is then transmitted to the Earth's surface by contact) results in a torque on the Earth's surface that corresponds to an extremely small acceleration of its surface and so also a distance moved in (say) one second – that *must* necessarily then lie in the quantum mechanical realm – this is just what is frustrating the usual classical mechanical analysis of the cart. So apparently the problem then is that one has exited the classical mechanical realm and entered the quantum mechanical realm ... which is then, in effect, just what Noll mentioned as a *key* problem in continuum mechanical calculations using classical mechanical systems of more and more particles at the beginning of his introduction that is reproduced in Appendix 6.

And, of course, as to the using of a scale to determine an object's inertial mass, when was the last time the Earth was put on a scale and weighed? Consequently, the author follows Prof. Noll's above mentioned idea in laying the difficulty of non-conservation of momentum at the door of the cart-earth system not being a suitable classical mechanical system because it then involves motions and hence distances that are out of the classical mechanical area of applicability; however, Noll was concerned with motions and hence distances that were too small while we are here concerned with motions and hence distances that appear to be too large for classical mechanical momentum conservation to apply.

# WHY DOES CLASSICAL MECHANICS FORBID INERTIAL PROPULSION DEVICES WHEN THEY EVIDENTLY DO EXIST?

# 3. The Making Of The Milkovic Cart Into A True Inertial Propulsion Device

We have mentioned at the beginning of our Introduction that the Milkovic cart is not – strictly speaking – an inertial propulsion device since it soon runs down and stops all motion. However, the author has been trying to think of a better example of a system of particles than his previous three particle example that Prof. Casey has pointed out is actually NOT at odds with the conservation of momentum; and now he thinks that he has finally found one; but it is much, much more complex this time. As the author mentions in the Introduction of this book, he has (successfully) computer simulated the Veljko Milkovic oblique pendulum driven cart for several swings of the pendulum back & forth (see his ResearchGate.com web page for this simulation that may be freely downloaded there). But Gottfried Gutsche points out that it is not a true IP device as it soon runs down, and then it halts, of course.

However, the author now sees how to correct this defect as follows: after one complete swing back & forth, the pendulum shaft connecting the oblique pendulum bob to the post (in the cart chassis) that lies in and moves in the oblique plane (having constant angle with respect to the cart post) is not horizontal as it was in starting out before swinging back & forth; but rather, then, the shaft is at a positive angle below the horizontal ... because of friction and the transfer of pendulum momentum to the chassis to move it (net) in the forward direction. So, then, at this exact moment when it completes its back & forth swinging with respect to the cart post; there is an added a mechanical device that grabs a rotating arm that is perpendicular to the shaft (at the post) and then applies a (fixed) torque to this arm so as to bring the shaft steadily up to the horizontal position, and finally then it lets the arm go free and the cycle begins again. Now, this arm that is perpendicular to the shaft is really one piece with the shaft so that it only rotates with respect to the pendulum's shaft as this shaft swings simply back & forth in its oblique plane relative to the cart. But, then, what effect does this lifting have on the cart's motion?

Well, since then the bob goes backwards in the oblique plane (that it swings in) relative to the cart post; the chassis tends, then, to simultaneously move forward. So, therefore, the observed (net) forward motion of the cart chassis during the single back & forth pendulum motion would only be augmented by this shaft & bob lifting at the end of one single back & forth cycle of swinging! Thus, we (finally) have a true IP device as it can continue cycling like this until the spring drive finally loses its tension and thus winds down.

Now it might be objected here that, on one hand, we claim that momentum is not generally conserved in (small) inertial propulsion devices with slowly moving parts; but, on the other hand, that in the just given case of a mechanical device lifting the cart bob at the end of each complete swing can be shown not to alter the cart's motion significantly due to what is basically a momentum conservation argument; but keep in mind that inertial propulsion is in all known cases (to the author) a rather complicated effect (this being the reason why it remained undiscovered so long) while the pendulum lifting is an un-complicated effect. It is the fact that the cart pivot point is the origin of an accelerated and hence non-inertial system that is the key to its behavior as an inertial propulsion device, and to its successful computer simulation as well.

# WHY DOES CLASSICAL MECHANICS FORBID INERTIAL PROPULSION DEVICES WHEN THEY EVIDENTLY DO EXIST?

# 4. An Extremely Elementary Mechanical Device Exhibiting Both Non-Conservation of Momentum and Also Non-Conservation of Energy

Recently, the work of (the late) Maurice Couloumbe has come to the attention of the author via Mr. Andre Michaud, Maurice's friend and translator, including his (2004) US patent of an inertial propulsion water-craft (see the end of Appendix 1 below). The author has excised one portion of the cyclic motion of an extremely elementary device of Couloumbe's (having only three parts that possess any significant inertial mass) which exhibits in its simple operation both non-conservation of momentum and also of energy, and that is *not* gravity machine as is the Milkovic oblique pendulum driven cart discussed in the three preceding chapters.

This was very surprising to the author as he had previously gone 100% with inventor and author, Gottfried Gutsche's and retired Boeing supervisor, Michael Gamble's assertions that although momentum and angular momentum are not separately conserved in an isolated mechanical system (as previously thought) having a finite number of particles all having positive inertial mass; that, never-the-less, energy (kinetic plus potential) was definitely conserved there. But "live & learn," one has to add. A diagram follows:

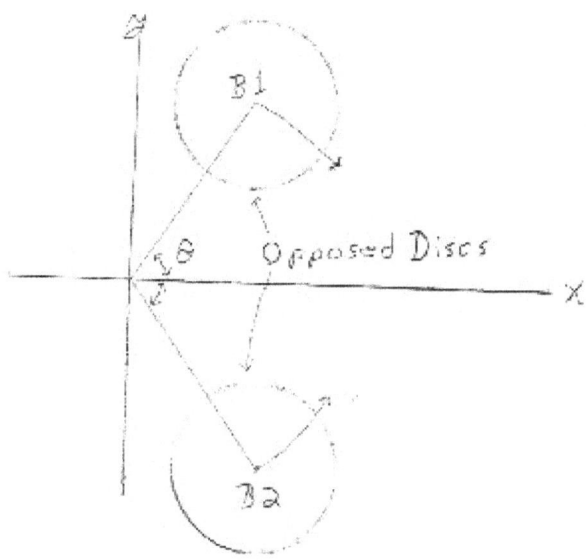

This device is shown to be essentially two dimensional (and so *not* a gravity machine as the x-y plane is assumed to be horizontal), and the two (opposed) discs might be best thought of a hockey pucks on wet & smooth ice having then very little friction ... as the device is taken to be completely friction-less. And, as the device is assumed to be symmetric with respect to the x-axis, it will move with its center of mass/pivot point along the x-axis in the positive x-axis direction as the two (equal)

angles "θ" go from 90 degrees to (almost) zero degrees due to centrifugal force from the rotating opposed discs on mass-less but rigid shafts pivoted at the common pivot point. The calculations of the motions of the opposing discs and also the pivot point and pivot mechanism (the latter being assumed having the pivot point as center of mass) are straight-forward classical mechanical ones ... except for the fact that the x-y system's origin is assumed to be at the pivot point and so is accelerated and thus not being an inertial system. (This is very similar to the Milkovic cart's key pivot point situation as identified at the end of Chapter 3 above.) But since we define inertia as "resistance to force" (following Newton), we could handle this non-inertial property by following (MIT) Prof. (Emeritus) A. P. French's widely used physics text, "Newtonian Mechanics" [5], but we choose to revise Prof. James Casey's solution instead. Then a system of two ordinary differential equations (each of second order) is obtained from Newton's second law of motion involving the two key time dependent parameters of the device; and then this system is numerically integrated using the Derive 5 mathematical computer language's Runga-Kutta numerical integrator. There are 53 time steps here, each of length $1/1000$ second from $t = 0$ to $t = 0.053$ second with the using of 10 significant digits precision in our calculations. And the results are quite clear: neither device momentum nor device kinetic energy are conserved over this time interval … even approximately. (See the author's ResearchGate.com web page for a preprint containing these two calculations in detail.) And since this device has been noted above not to be a gravity machine and thus because of its symmetry involving the two opposed discs so that the cart — on which it is assumed to be mounted -- exerts no sheer force on the straight and level tracks, then it should also work the same in (say) outer space far from any gravitating mass as well. One must conclude, then, that the well-known prohibition of any "over unity" free energy mechanical device because of (postulated) conservation of energy (in an isolated system) must be sent to the trash bin of history along with, for example, the phlogiston theory of heat.

This is accomplished by using a black box model of the entire device that then leads to a variable inertial mass for the interior of the box and then we introduce a concept of mechanical dependence in analogy with dependence of probabilistic events, namely, that their joint probability cannot be calculated by simply multiplying their standalone individual probabilities but the exact dependence must be obtained by careful analysis. In much the same way, the Couloumbe device's dynamics cannot be exactly obtained by the usual method of analyzing the forces and fixed masses, writing this in equations, and then solving these equations mathematically, but rather the above mentioned black box technique involving device variable inertial mass needs to be employed.

# WHY DOES CLASSICAL MECHANICS FORBID INERTIAL PROPULSION DEVICES WHEN THEY EVIDENTLY DO EXIST?

# 5. A Further Analysis of the Couloumbe Device

But now the question arises: why is it that this device fails to exhibit non-conservation of momentum? Since it is not a gravity machine and so would be expected to operate also in outer space far from any gravitating matter, one certainly cannot blame any supposed change in the Earth's center of mass here as being way too small.

Well, it's already been pointed out above that it seems to be the fact that the Milkovic cart's analysis and the Couloumbe device's analysis both have natural coordinate systems that are accelerated which is (somehow) the key to this problem, because it must be pointed out that Lagrangian methods might be successfully applied to the Couloumbe device as it has no friction to cause trouble; but C. Lanczos points out on pages 403-4 of his (Dover book) classic, "The Variational Principles of Mechanics," that the use of Lagrangian mechanics implies conservation of momentum an also of energy (due to some work of Emmy Noether). But he also mentions in this book that should Lagrangian methods fail in ordinary low velocity situations due to, for example, friction; then one has to revert back to Newton's mechanics instead ... wherein we have already noted that momentum need not be conserved. Thus, it would appear that our Newtonian result that momentum need not be conserved carries more weight than any Lagrangian mechanical solution to the motion of the Couloumbe device in which momentum – according to Lanczos – will then be conserved ... *automatically*. Moreover, on pages 76-7 of Lanczos's book, he discusses the physical interpretation of the principle of virtual work; and he formulates this key principle as "Proposition A: The virtual work of the forces of reaction is always zero for any virtual displacement which is in harmony with the given kinematic constraints" which he then mentions also holds for dynamics as well as statics and so then additionally for all analytic dynamics. And, in a footnote to this discussion, he also mentions that although some scientists claim that analytical dynamics follows from the laws of Newton, he cannot verify this himself. Hence, it would seem that the possibility of non-conservation of both momentum and also of energy in Newton's mechanics need not necessarily run counter to their proven (by Emmy Noether) conservation in analytic dynamics as the latter fails to follow from the former.

And, as to the Milkovic cart, the conservation of

momentum could fail for more than just one good reason, it would certainly seem. And, of course, it is the non-conservation of momentum that then allows the two devices to exhibit inertial propulsion in their prescribed motions.

But, then, just exactly what goes wrong with the usual Newton mechanical proof of conservation of momentum for the Couloumbe device? Well, again, it appears that a key factor is the third law of motion which French notes that gives slightly different types of forces than his second law (and H. Hertz also notices this in his classic book on mechanics in a new form), and so the usual short proof of momentum conservation needs to be carefully re-examined regarding to this fact in order to avoid just basically saying "forces are forces are forces," and then continuing on one's way as if that settled matters completely. Because it evidently does not settle matters in a system of particles involving a small mechanical device such as the Couloumbe device, where we will also see in some of the appendices below that there actually exists a whole theory of inertial propulsion devices involving Boeing and other lesser-known aero-businesses (such as Gutsche's).

# WHY DOES CLASSICAL MECHANICS FORBID INERTIAL PROPULSION DEVICES WHEN THEY EVIDENTLY DO EXIST?

# APPENDIX 1
# A SUMMARY OF TECHNIQUE

We now summarize our ideas on how to proceed mechanically so as to avoid calculation problems, where **we assume that variable inertial mass** – treated in Dr. Jeremy Dunning-Davies's and our [2] – **does not occur** to any significant degree:

(1) While it's not true, in general, that either momentum or angular momentum are separately conserved, and so these two "laws" should never be employed, in general; still, in many special cases, they are valid. However, it's best not to use either; and to instead to use equations closely tied to the physics of the devices or situations that are under analysis.

(2) In ordinary (low-tech) mechanical calculations, not involving electromagnetic phenomena, energy does seem *often* to be conserved with kinetic energy being the usual ($\frac{1}{2}$ m v$^2$) and gravitational potential energy being the usual (m g h), and so energy methods may be used *cautiously* to obtain solutions in closed form. (But should variable inertial mass occur significantly, then the definition of kinetic energy must be *uniquely* altered to retain work-energy equivalence [2, Chapter 4 Appendix].)

(3) However, in case the solution is not to be or cannot be obtained in closed form, but must instead be

obtained by numerically integrating the system of ordinary differential equations of motion with the independent variable being time; then it is best not to use energy methods since then numerical accuracy problems may arise. Our preliminary work seems to indicate that since we accept the notion of Prof. Oleg D. Jefimenko's that the cause must precede the effect in physics (see Appendices 2 and 3 below), then it follows that causality propagates through a system at finite velocity. So, then, since there is kinetic heat in the wheel bearings and with this heat propagating generally much slower than the both the pendulum bob and the chassis velocities, it seems that this propagation disparity may well explain the poor results numerical integration yields when energy methods are utilized.

(4) So, therefore, since in a case where angular momentum, moments of inertia, and so on ... are used, then often conservation of energy must also be used to obtain the correct answer – such as in a nutating gyroscope with a fixed pivot point – where angular momentum methods fail without the additional use of energy methods; it is best also not to use such rotational physics if there is going to be the numerical integration of equations of motion in view of (3) just above. And this may seem almost impossible, but since angular momentum physics is derived from (linear) momentum physics [1, 3, 4],

this is always possible; and it also seems to be always possible to avoid energy methods too ... as energy methods are also derived from momentum methods ... although the use of energy methods will tend to make it *much* easier to obtain closed form solutions to the (ordinary differential) equations of motion ... if this is actually possible in the reader's problem.

(5) To illustrate the above, we direct the attention of the reader to the author's computer simulation of the Veljko Milkovic oblique pendulum driven cart inertial propulsion device (mentioned above in our Introduction) on the research web site "ResearchGate" ... where it may be downloaded for no cost under the author's name there. And we also invite the reader to look there at our computer simulation of the Couloumbe device which is both considerably more elementary and not a gravity machine as well.

Now, with the above five pointes firmly in mind, we might now give an analogy with surveying to illustrate our thinking on this subject. Our analogy is with surveying on the surface of the Earth where there are no hills or valleys. For a such a small plot of land, a surveyor can then ignore the fact that the Earth taken as a whole is approximately spherical since the Earth's radius is quite large, and so his surveying then only requires a knowledge of plane geometry and plane trigonometry, But, if the plot of land to be surveyed is too large, he will not be able to get the required accuracy of measurement this way, and so then he must resort to the more complicated spherical geometry and spherical trigonometry instead.

In much the same way, we strongly suspect that the usual proof of conservation of momentum for small devices in classical mechanics goes wrong when applied to a device that is too large in complexity. However, as long as the device's mechanics is not too complicated, ordinary conservation of momentum may be prudently used. But it is evidenced, for example, in an elementary watercraft inertial propulsion device [5] patented by (the late) Maurice Couloumbe of Quebec, Canada, that this complexity limit need not be all that large ... as is discussed in our Chapters 4 and 5 above.

### References

[1]   G. E. Hay, *Vector and Tensor Analysis*, pp 66-101 (Dover, 1953).

[2]   Dennis P. Allen Jr., and Jeremy Dunning-Davies, *Neo-Newtonian Mechanics With Extension to Relativistic Velocities; Part 1: Non-Radiative Effects,* Ninth Edition, (CreateSpace.com, 2018).

[3]   J. L. Synge & B. A. Griffith, *Principles of Mechanics* (McGraw–Hill Book Company, 1949).

[4]   A. P. French, *Newtonian Mechanics,* (W.W. Norton & Co., 1971).

[5]   M. Couloumbe (2004) U.S. Patent No. 6,716,074 B2 Washington DC: U.S. PATENT AND TRADEMARK OFFICE.

# APPENDIX 2

# CAUSALITY IN MECHANICS

Dennis P. Allen Jr.

There appears to be considerable confusion in classical physics, not involving electromagnetic or gravitational phenomena, concerning causality. The late Prof. Oleg D. Jefimenko writes near the beginning of Chapter 1 of his "Causality Electromagnetic Induction and Gravitation" that: "One of the most important tasks of physics is to establish causal relations between physical phenomena. No physical theory can be complete unless it provides a clear statement and description of causal links involved in the phenomena encompassed by that theory. In establishing and describing causal relations it is important not to confuse equations which we call 'basic laws' with 'causal equations.' A 'basic law' is an equation (or a system of equations) from which we can derive most (hopefully all) possible correlations between the various quantities involved in a particular group of phenomena subject to this 'basic law.' A 'causal equation,' on the other hand, is an equation that unambiguously relates a quantity representing an effect to one or more quantities representing the cause of this effect. Clearly, a 'basic law' need not constitute a causal relation, and an equation depicting a causal relation may not necessarily be among the 'basic laws' in the above sense."

"Causal relations between phenomena are governed by the *principle of causality*. According to this principle, all present phenomena are exclusively determined by past events. Therefore, equations depicting causal relations between physical phenomena must, in general, be equations where a present-time quantity (the effect) relates to one or more quantities (causes) that existed at some previous time. An exception to this rule are equations constituting causal relations by definition; for example, if force is defined as the cause of acceleration, then the equation $\mathbf{F} = m\mathbf{a}$, where $\mathbf{F}$ is the force and $\mathbf{a}$ is the acceleration, is a causal equation by definition."

"In general, then, according to the principle of causality, an equation between two or more quantities simultaneous in time but separated in space cannot represent a causal relation between these quantities because, according to this principle, the cause *must precede* its effect. Therefore the only kind of equations representing causal relations between physical quantities, other than equations representing cause and effect by definition, must be equations involving 'retarded' (previous-time) quantities."

It is evident that Jefimenko sees no way to introduce causality into mechanics other than by definition. And Prof. A.P. French, in his widely used "Newtonian Mechanics" beginning physics text, also appears to be similarly confused as he says in his section on gyroscopic nutation; "However convincing the analysis of gyroscopic precession may seem, one may still wonder how a gyroscope can possibly defy gravity in the way it appears to do. The answer is that this immunity *is* indeed only apparent. If a flywheel is set spinning about a horizontal axis, with both ends of the axle supported, the first thing that happens if the support at one end (A) is removed is that this end does begin to fall vertically. Immediately thereafter, however, the precessional motion in a horizontal plane begins, and as this happens the falling motion slows down, until the point A is moving in a purely horizontal direction. It does not stay like this; what happens next is that the precession slows down and the end of the axle rises again, ideally to its initial level. The whole sequence is repeated over and over ... The process is called *nutation*..."

Thus French also seems to fall short of demonstrating causality ... although he seems to allude to the idea that first in this gyroscopic situation (after the gyroscope at $t = 0$ suddenly becomes unsupported at one end) nutation begins which then immediately causes precession to commence – a sort of causality that is apparently not completely definitial as in Jefimenko's just given quotation; but the difficulty is that this simultaneity is shown by the exact solution to the system of two second order ordinary differential equations describing the ensuing precession and so on. However, this difficulty is easily obviated as follows:

First notice that empirical physics has the property that since measurement of physical variables is only approximate to just so many significant figures, this means mathematically that one begins by "making the continuum discrete" in that (say) the relevant physical variables can only be measured to one significant figure, then if we truncate (rounding is much the same) our numbers in (for example) French's nutation case (just quoted), then all numbers x with $2 \leq x < 3$ will then assigned the one significant figure 2 ... and so on. [In the case of (say) $0 < x < 1$, we note that if we write x scientifically as (k $10^n$), then clearly the absolute value of n is bounded in our experimental work.] Thus, when we assign measured numbers to this gyro situation and then numerically integrate the system of two second order ODE's (while it may appear that French has one first order and one second order ODE, nevertheless, just above the first numbered first order ODE is the second order ODE it came from via integration) by Euler's method (the most elementary and straight-forward method) [1] after choosing a sufficiently small time step $\Delta t > 0$; instead of referring to French's solution, we see that the nutation angle (measured from the horizontal) together with its time derivative and also the precession angle together with its first two time derivatives are all zero at $t = 0$ (the initial conditions); but when the one support is removed, nevertheless, the second nutation angle time derivative does **not** vanish as it is accelerated by gravity instantly. This results in the initial values of all but the second time derivative of the nutation vanishing at $t = 0$, but after a time step of $\Delta t$, we see that the first time derivative of the nutation then also becomes non-zero, and, of course, the nutation second time derivative remains non-zero too as a time step of $\Delta t$ occurs ... and the precession second time derivative may now become non-zero too after this one step. But the other three quantities remain zero here. Further, after another such time step, the nutation angle then becomes non-zero too, just as the nutation first and second time derivatives are non-zero as well. However, what about the precession angle? We find that the precession angle is still zero after two time steps ... although the nutation angle is not! Thus, in making the continuum discrete, one sees here that the nutation precedes the precession, and so it can then be said in the sense of Jefimenko above that there **is** a true causal relation here with the nutation causing the precession as the physical process develops from $t = 0$!

It should be noted that the continuum is dearly beloved by mathematicians, and even the late Prof. Errett Bishop, in his monumental "Foundations of Constructive Analysis," mentions that Luitzen Brower (of the Brower fixed point theorem and an important earlier constructivist as well as one of the founders of modern topology) seemed to feel that the continuum would [constructively] turn out to be discrete "if he did not personally intervene"! But continuum mathematics, nevertheless, obscures causality in mechanics, and that is rather unfortunate, of course! This clearly illustrates that the over-mathematization of physics nowadays is certainly not without its deleterious foundational effects!

Finally, we heartily recommend (the late) Prof. Robert M. Kiehn's six volumes in "Non-Equilibrium Systems and Irreversible Processes" … as he, too, has investigated the possibility that continuum mathematics might not always be the right setting for theoretical physics … and very extensively as well in that he, for example, completely disregards the traditional point particle approximation approach and works with *finite* topologies that he then characterizes completely.

## References

[1] Wilfred Kaplan, *Ordinary Differential Equations,* 400-1, Addison-Wesley Publishing, 1958; (but the author does not mention Euler's name).

# WHY DOES CLASSICAL MECHANICS FORBID INERTIAL PROPULSION DEVICES WHEN THEY EVIDENTLY DO EXIST?

## APPENDIX 3

## A NUMERICAL CALCULATION TO ILLUSTRATE THE PREVIOUS APPENDIX'S DESCRIPTION OF THE EULER INTEGRATION OF A GYROSCOPE'S TWO SYSTEM OF ORDINARY DIFFERENTIAL EQUATIONS SHOWING CAUSALITY

We refer the interested reader to the web site "ResearchGate.com" where, under the author's name, there is a spread-sheet (that may be freely downloaded) containing an detailed (numerical) Euler integration of the system of two ordinary differential equation found in A. P. French's "Newtonian Mechanics"[5] under the heading of gyroscope nutation. This numerically illustrates the words of the previous appendix.

# WHY DOES CLASSICAL MECHANICS FORBID INERTIAL PROPULSION DEVICES WHEN THEY EVIDENTLY DO EXIST?

# APPENDIX 4

## BOEING'S LONG HISTORY OF USING INERTIAL PROPULSION TO REPOSITION THEIR SATELLITES INTO NEW AND DIFFERENT ORBITS

Long time Boeing engineering supervisor, Michael Gamble, has given a talk recently at the "Seventh International Conference On Future Energy" (COFE7) concerning the extensive history of Boeing's using inertial propulsion (IP) of the forced precession type ... that still continues today. (He refers to this in his talk by using the company name "Control Moment Gyros".) And a DVD of this COFE7 talk may be ordered from the conference sponsor, the "Integrity Research Institute" at (888) 802-5243.

This is especially significant because there is, of course, very little air friction in outer space, and such IP devices as discussed in our introduction are usually attempted to be explained away by naysayers as frictional effects of some sort. But, needless to say, such arguments cannot and do not apply to Boeing's multimillion dollar IP technologies as exposited by Gamble.

His talk is about a hour in length, and he is quickly seen to be a good, solid engineer whose explanations are both clear and concise. There is **no** ambiguity nor any esoteric theory in his presentation ... that also contains many, high quality photographs of the actual equipment used by Boeing over the years.

The author highly recommends the DVD containing

Gamble's talk to the interested reader!

Very recently in COFE9, Gamble has discussed a table-top model of the Boeing CMG inertial propulsion technology that he and Dr. Thomas Valone have developed and carefully tested, and the results of which hopefully will soon be written up and published in a reputable journal so as to (finally) make the mechanical engineering profession aware of the reality of inertial propulsion.

But many, many scientists and engineers today have simply rejected this talk *out of hand* ... because they are *absolutely certain* that were mechanical inertial propulsion actually possible (that, of course, violates conservation of momentum by definition), then it certainly would have been discovered long, long ago with all the high tech advances in (say) the last 100 years; however, the author has a recent book, "Foundations of Gutschian Mechanics; Part 1: Basics" in which the new energy mechanics of Mr. Gottfried Gutsche is expounded upon that hopefully will open up this mechanics of Gutsche's which allows him to design and build mechanical inertial propulsion devices which he then patents and sells. And so, of course, his mechanics does not embrace the separate conservation of angular momentum and momentum of classical mechanics; but rather allows for the conversion of angular momentum to momentum (and vice versa) in certain important cases of interest. But Gutsche is not a native English speaker, and his books on IP have proven quite opaque to his potential readers including the mainstream engineers and physicists. And so this new book of the author is designed to be the key to unlock Gutsche's mechanical thinking.

# APPENDIX 5

# A BRIEF INTRODUCTION TO GOTTFRIED GUTSCHE'S POINT OF VIEW IN HIS INERTIAL PROPULSION WRITINGS

In this appendix, we aim to introduce Gutsche's point of view in his inertial propulsion devices. It centers on the analysis of the flow of mechanical energy within mechanical devices ... that begins with potential energy (for example, a compressed spring) and then flows from this. And his key simple device plays a similar role in his theory to the simple harmonic oscillator's role in classical mechanics.

This device is a pair of masses that are not, in general, the same; but they are located at opposite ends of a simple coiled Hooke type (massless) spring, and are allowed to oscillate freely & without any friction. Thus, if the spring between the two masses M and m is compressed and then released at t = 0, the device's subsequent oscillations are tracked by the laboratory velocities V and v of M and m, respectively. One certainly *could* analyze the device's compound motion using the conservation of momentum applied to the system's center of mass that might be taken as the center of coordinates, but this is quite unnecessary as Newton's second and third laws suffice without the conservation of momentum applied to the center of mass's velocity vector. And, in fact, one need not even employ the definition of the center of mass of a system of particles at

all here!

Now, Gutsche's key insight is that at any time $t > 0$, we have $M/m = E/e$, where $e = \frac{1}{2} M v^2$ and $E = \frac{1}{2} m V^2$, the kinetic energies of masses M and m, respectively. That is, although in Newton's theory, mass gravitates toward other mass with a larger mass resulting in a proportionally stronger attraction; yet, in the case of mechanical kinetic energy, this energy moves rather toward smaller mass concentrations and away from larger mass concentrations ... if it is free to flow or move ... as it is in his key simple two masses and a spring device. Thus, a bullet fired from a (more) massive gun will be found to receive more kinetic energy than the gun, itself.

Then he goes on to introduce a new mechanical concept called (by him) the "mechanical kinetic energy momentum" and having formula ($\frac{1}{2} m^2 v^2$) ... that may helpfully be viewed either as half the dot product of the momentum vector with itself ... or else as a simple product of the mass and the kinetic energy ... so that, in this key simple device, both masses at any time $t > 0$ have equal mechanical kinetic energy momentums.

Thus, this novel concept may be viewed as a "hybrid" concept lying between the momentum and the kinetic energy, and the derivative of this mechanical kinetic energy momentum with respect to the scalar momentum is just this scalar momentum itself.

This, then, leads him to proclaim that "momentum is conserved *in the energy form* of kinetic energy".

Now, in electrodynamics, it was John Henry Poynting who originated the "Poynting vector" that is the key to tracking the flow of electromagnetic energy and also of electromagnetic momentum … with the two being connected by Einstein's famous $E = M c^2$ (that, however, was known earlier to J. J. Thompson). But, to the best of the author's knowledge, the topic of energy and momentum flow in mechanical devices is not usually treated in the best mechanical and dynamical texts very extensively … as, however, it certainly is in the best electrodynamic texts (see [9], for example) with Poynting's theorem and all.

More recently, Gutsche has published: *Inertial Propulsion; the internal frequency modulated mechanical oscillator* [6] that further involves the idea of Eric Laithwaite, an electrical engineer (as also is Gutsche) who aimed to use alternating current circuital concepts in order to study the newly discovered (by him and Alex Jones before him) concept of variable inertial mass and inertial propulsion, and so as to suggest a whole new type of mechanics involving these electrical dynamical notions.

Additionally, it should be noted that Gutsche – like Prof. Robert M. Kiehn before him (see Appendix 2 above) – does not go 100% with continuum mathematics and, in fact, does not utilize Newton's familiar "F = m a" to obtain his inertial forces [4].

The author hopes and prays that this brief introduction to Gutsche's rather unorthodox mechanical thinking …

and especially to his very original inertial propulsion ideas … will prove helpful to the reader in understanding his convoluted inertial propulsion device writings that have proven so very opaque to so many of his prospective readers!

Finally, it should be mentioned that the author has a new book, "Foundations of Gutschian Mechanics; Part 1: Basics" where he expounds upon Gutsche's mechanical thinking.

# WHY DOES CLASSICAL MECHANICS FORBID INERTIAL PROPULSION DEVICES WHEN THEY EVIDENTLY DO EXIST?

# APPENDIX 6

*Symposium on the Axiomatic Method*

## THE FOUNDATIONS OF CLASSICAL MECHANICS
## IN THE LIGHT OF RECENT ADVANCES
## IN CONTINUUM MECHANICS [1]

WALTER NOLL

*Carnegie Institute of Technology, Pittsburgh, Pennsylvania, U.S.A.*

1. **Introduction.** It is a widespread belief even today that classical mechanics is a dead subject, that its foundations were made clear long ago, and that all that remains to be done is to solve special problems. This is not so. It is true that the mechanics of systems of a finite number of mass points has been on a sufficiently rigorous basis since Newton. Many textbooks on theoretical mechanics dismiss continuous bodies with the remark that they can be regarded as the limiting case of a particle system with an increasing number of particles. They cannot. The erroneous belief that they can had the unfortunate effect that no serious attempt was made for a long period to put classical continuum mechanics on a rigorous axiomatic basis. Only the recent advances in the theory of materials other than perfect fluids and linearly elastic solids have revived the interest in the foundations of classical mechanics. A clarification of these foundations is of importance also for the following reason. It is known that continuous matter is really made up of elementary particles. The basic laws governing the elementary particles are those of quantum mechanics. The science that provides the link between these basic laws and the laws describing the behavior of gross matter is statistical mechanics. At the present time this link is quite weak, partly because the mathematical difficulties are formidable, and partly because the basic laws themselves are not yet completely clear. A rigorous theory of continuum mechanics would give at least some precise information on what kind of gross behavior the basic laws ought to predict.

I want to give here a brief outline of an axiomatic scheme for continuum mechanics, and I shall attempt to introduce the same level of rigor and clarity as is now customary in pure mathematics. The mathematical

[1] The results presented in this paper were obtained in the course of research sponsored by the U.S. Air Force Office of Scientific Research under contract no. AF 18 (600)–1138 with Carnegie Institute of Technology.

# WHY DOES CLASSICAL MECHANICS FORBID INERTIAL PROPULSION DEVICES WHEN THEY EVIDENTLY DO EXIST?

# APPENDIX 7

We consider Veljko Milkovic's oblique pendulum driven cart inertial propulsion device (see his interesting web site) via classical mechanics, but without using ANY energy methods or angular momentum methods whatsoever. The device (see following diagrams) is an oblique (or slanting forward) pendulum mounted upon a chassis that is free only to roll forward or back so that if the x-y plane is horizontal and non-rotating, then the pendulum slants forward making a negative acute angle of $\alpha$ radians (here negative because the plane slopes downward as x becomes larger) and the plane contains the y-axis of the x-y-z Cartesian coordinate system which moves with the device pivot point that is fixed at the system origin. The slant angle $\alpha$ remains fixed in time, but the device chassis moves in a straight line ... and it has coordinate X (so that $DX = dX/dt = v$) that vanishes at $t = 0$. The velocity vector of the pendulum bob with respect to its pivot point that is fixed on the chassis has (in the x-y-z system) coordinates $[Dx, Dy, Dz]$ with respect to the x-y-z coordinate system whose origin is at the (moving) pendulum pivot point. The swing angle $\theta$ is negative to the left of the x-axis and positive to the right of the x-axis when viewed from the positive values of this axis.

The bob at $t = 0$ is at swing angle $\theta = -\pi/2$ radians, and when allowed to swing freely, it swings downward in the oblique plane due to gravity and it also causes the plane to parallel translate due to the pendulum sloping downward through a (negative) acute angle $\alpha$.

See the diagrams below (due to Gottfried Gutsche):

# WHY DOES CLASSICAL MECHANICS FORBID INERTIAL PROPULSION DEVICES WHEN THEY EVIDENTLY DO EXIST?

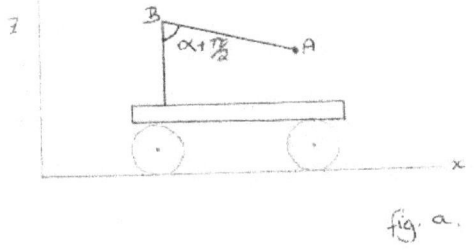

fig. a.

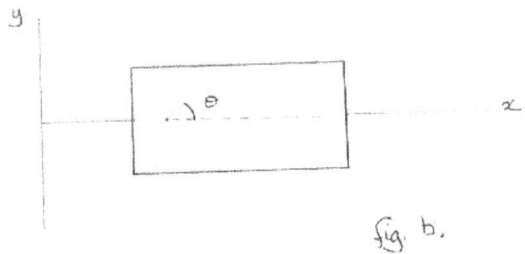

fig. b.

# REFERENCES

[1] Gottfried J. Gutsche, *Inertial Propulsion: the quest for thrust from within,* (CreateSpace.com, 2014).

[2] Dennis P. Allen Jr. & Jeremy Dunning-Davies, *Neo-Newtonian Mechanics with Extension to Relativistic Velocities; Part 1: Non-Radiative Effects,* Ninth Ed., (CreateSpace.com, 2018).

[3] Harvey E. Fiala, "An Inertial Propulsion Patient & Working Model", Presentation, July 29, 2012, Tesla Tech, Inc., Marriot Pyramid North, Albuquerque, New Mexico.

[4] Gottfried J, Gutsche, *Inertial Propulsion; And You Thought It's Impossible,* (CreateSpace.com, 2018).

[5] A. P. French, *Newtonian Mechanics,* (W.W. Norton & Company, 1971).

[6] Gottfried J. Gutsche, *Inertial Propulsion; the internal frequency modulated mechanical oscillator,* (CreateSpace.com, 2017).

[7] Walter Noll, *The Foundations Of The Classical Mechanics In The Light Of Recent Advances In Continuum Mechanics,* Symposium on the Axiomatic Method, Univ. of California at Berkeley (1957).

[8] Alexander L. Dmitriev, *Frequency Dependence of Rotor's Free Falling Acceleration and Inequality of Inertial and Gravity Masses,* http://arXiv.org.1101.4678v1 [physics.gen-ph], pdf(2011)

[9] J. D. Jackson, *Classical Electrodynamics,* Third Ed., (Wiley, 2009).

[10] H. Lamb, *Dynamics,* (Cambridge University Press, 1961).

## ABOUT THE AUTHOR

The author earned his doctorate, master's and bachelor's degrees from the University of California at Berkeley in mathematics. He has done research work for Bell Telephone Laboratories and taught mathematics at Michigan Technological University. And he has written on gravitation as an electrical phenomenon and its application to earthquake early warning, and co-authored two books with Dr. Jeremy Dunning-Davies with one on a minimal mechanics that takes into account inertial mass change that has definitely been experimentally detected by scientists such as Harvey Fiala, a retired Space Shuttle Engineering Reentry Supervisor and master's degree student of Prof. Richard Feynman at Cal Tech, and the late Eric Laithwaite, inventor of the trains in Germany and Japan that float on magnetic fields and so do not touch the rails, and one on Ernst Mach type effects as well as on the recently discovered vector, not (GRT) tensor, gravitational waves that have been experimentally generated (in a vacuum) and then studied by Russian Prof. V.N. Samohkvalov. Moreover, he has authored a book on the Lebesgue measure that touches upon the continuum problem utilizing his work on the fundamental & primary cause of the first digit phenomenon and also his thesis work in algebraic automata theory under Prof. (Emeritus) John L. Rhodes (Univ. of Calif. at Berkeley's Mathematics Dept.). And he has a brief memoir in which he gives his experience in science and academia ... as well as his philosophy of science and of truth in general. All of his books may be found on Amazon.com where they may be viewed electronically.

# WHY DOES CLASSICAL MECHANICS FORBID INERTIAL PROPULSION DEVICES WHEN THEY EVIDENTLY DO EXIST?